HELLO

MY NAME IS

ROCK
POINT

The moon is the Earth's only

satellite and was formed 4.6 billion years

The impulse engines masterfully used on Star Trek were based on **F Ü Si O N** reactions. In reality, we rely on chemical fueled rockets, but NASA...

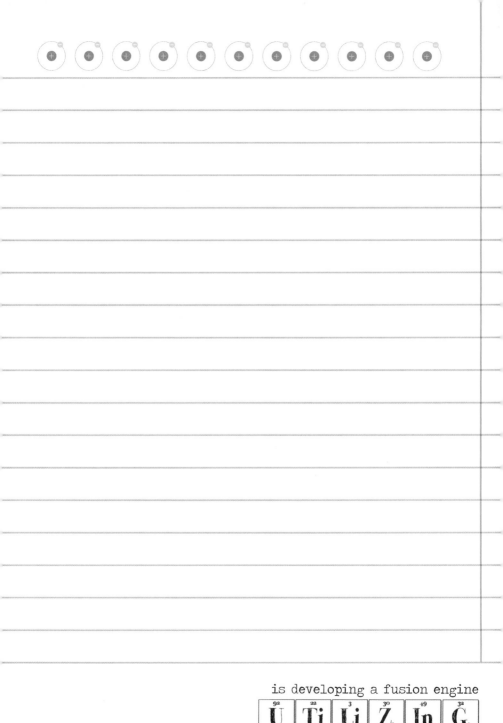

is developing a fusion engine

| ^{92} U Uranium 238.0 | ^{22} Ti Titanium 47.87 | ^{3} Li Lithium 6.938 | ^{30} Zn Zinc 65.38 | ^{49} In Indium 114.8 | ^{32} Ge Germanium 72.63 |

plasma created from deuterium and tritium. Go Scotty!

The Eiffel Tower

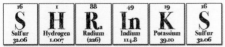

16	1	88	49	19	16
S	**H**	**R**ₐ	**In**	**K**	**S**
Sulfur	Hydrogen	Radium	Indium	Potassium	Sulfur
32.06	1.007	(226)	114.8	39.10	32.06

and expands an average of...

6 inches per year due to temperature changes and

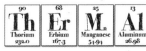

expansion.

ROYGBIV is how scientists remember the orders of the

in the electromagnetic spectrum.

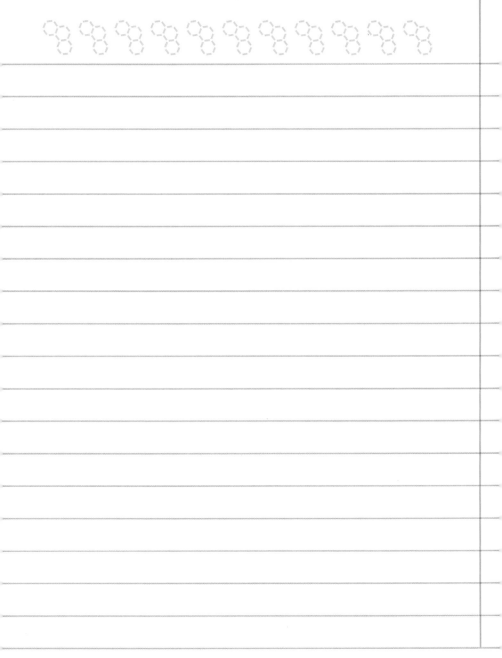

Blue light has

energy, a shorter wavelength and a
higher frequency than red light.

How close are we to meeting aliens?

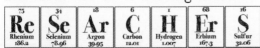

| 75 Re Rhenium 186.2 | 34 Se Selenium 78.96 | 18 Ar Argon 39.95 | 6 C Carbon 12.01 | 1 H Hydrogen 1.007 | 68 Er Erbium 167.3 | 16 S Sulfur 32.06 |

are using infrared mapping throughout our Universe...

trying to find "heat waste" that could indicate
complex biological life forms converting stellar

into useable forms of energy.

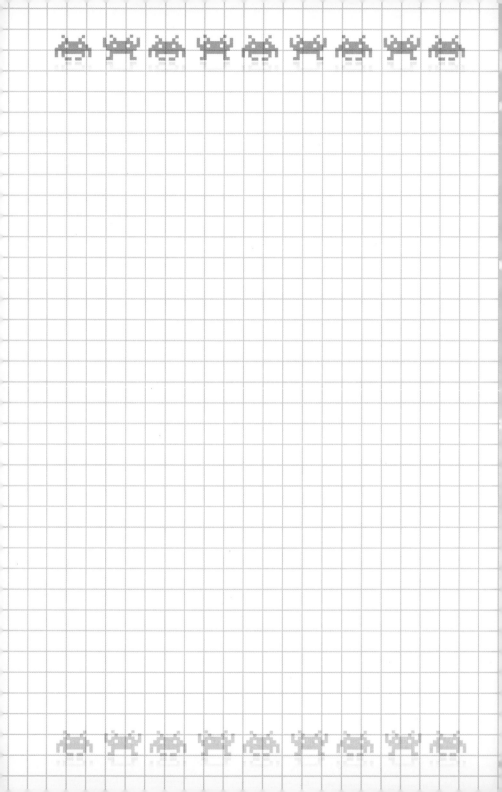

Why do we weigh less on the **Mo** **O** **N** and more on Jupiter? Because weight is a quantified reflection of your mass, the amount of matter in your body,...

multiplied by the gravitational force of the you inhabit.

Atoms are over 99.9% empty

16	91	58
S	**Pa**	**Ce**
Sulfufr	Protactinium	Cerium
32.06	231.0	140.1

Because of this, if you could squeeze all of the empty
space out of all the atoms…

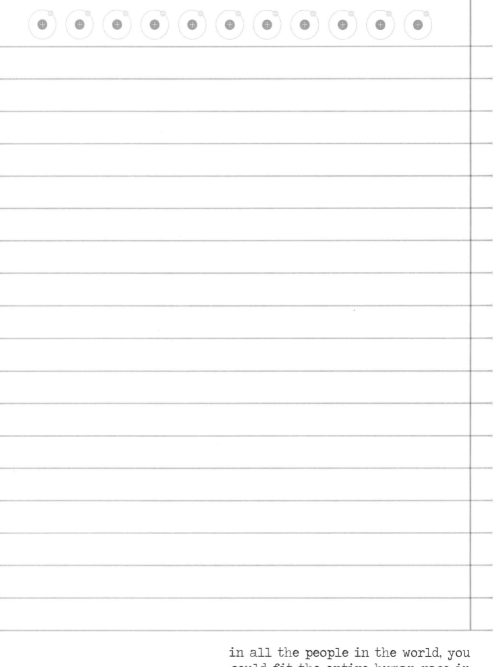

in all the people in the world, you
could fit the entire human race in
the volume of a sugar

What do you call iron blowing in the | **W** Tungsten 183.8 | **I** Iodine 126.9 | **Nd** Neodymium 144.2 | ?

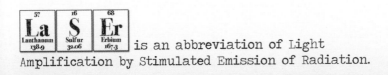 is an abbreviation of Light
Amplification by Stimulated Emission of Radiation.

Noble gas **Xe** **No** **N** lasers can cut through materials so tough even diamond tipped blades will not cut them.

Temperature measures the average

energy of the molecules.

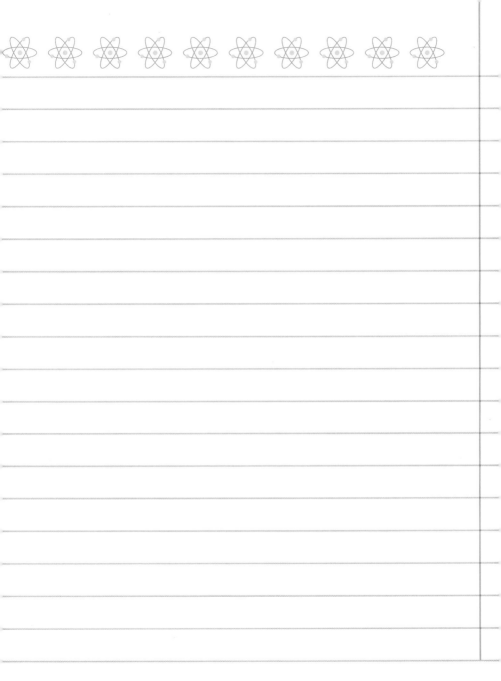

energy always flows from an object at higher
temperature to one of lower temperature.

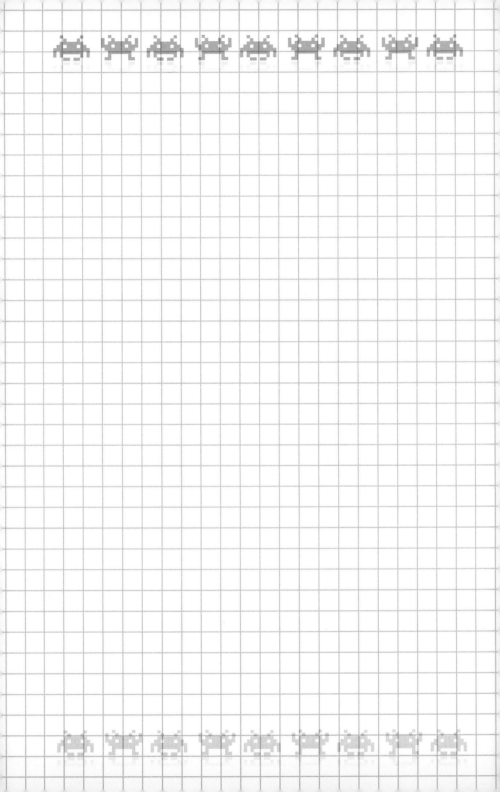

Because of the effects of gravity, when the MoON

is directly overhead,

39	8	92
Y	**O**	**U**
Yttrium	Oxygen	Uranium
88.91	16.00	238.0

weigh slightly less.

Al **E**s **X**e **A**u **Nd** **Er** Graham Bell created a form of electric telegraph that sent signals as musical notes made by vibrating reeds.

This idea led him to invent a way of

and receiving the frequencies present in
the human voice… the telephone.

Sound travels nearly five times faster in

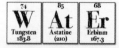

than in air and almost 20 times…

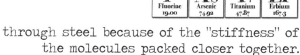

through steel because of the "stiffness" of
the molecules packed closer together.

The **E**arth's moon has changed very little in millions of years, due to its lack of atmosphere and weathering effects.

8 8 8 8 8 8 8 8 8 8 8

12 astronauts have been on the moon and over 880 lbs.
of moon rock have been brought back to Earth for

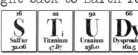

Spectrometers **Co** **N** **Ta** **In** prisms that split light from a star into a spectrum that can be analyzed.

Astronomer Annie Jump Cannon sorted thousands of stars
based on their unique spectra, which determined their

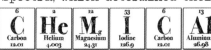

composition.

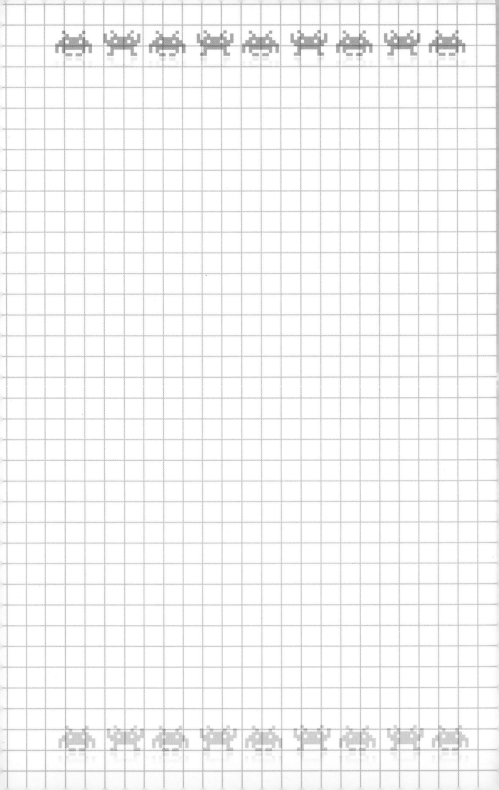

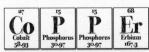

metal and its alloys, containing at least 60% copper...

are naturally anti-

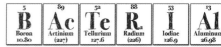

The amount of **Ca Rb O N** in the human body is enough to fill about 9,000 "lead" pencils.

An average adult contains around 250g (1/2lb) of salt.

An  of gold can be stretched
into a wire 80 kms (50 miles) long...

and absolutely **P U Re** gold is so soft it can be molded with the hands.

did the chicken cross the Mobius strip?

The air around humans is made up primarily of
Nitrogen (almost 80%). Bags of potato

are filled with nitrogen to

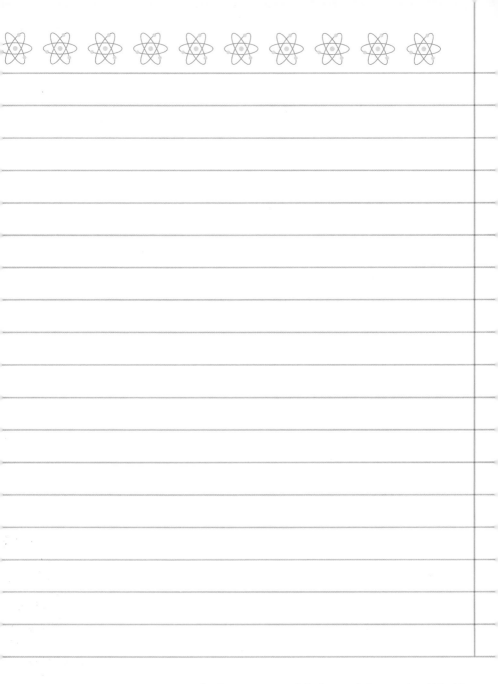

exclude oxygen which would react with the

and make them go stale.

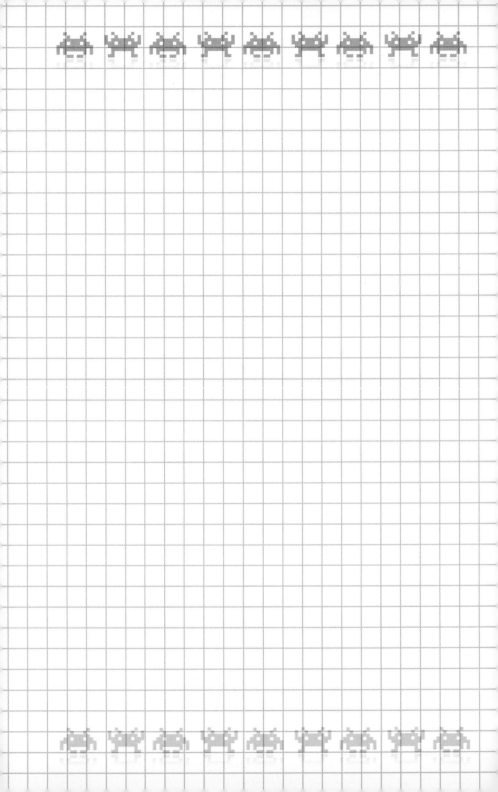

At **Te** **N** different times in the past three
million years, the Earth's magnetic field has flipped...

reversing itself. Scientists aren't

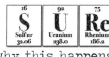

why this happens.

Most cars use unleaded gasoline, because lead can prevent platinum and rhodium

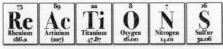

from occurring

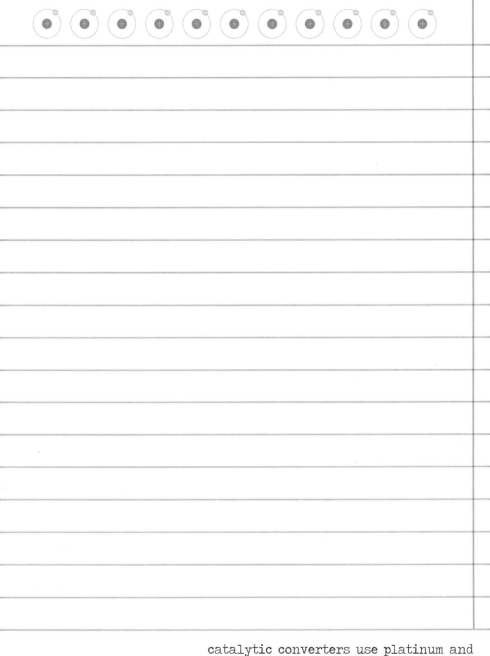

catalytic converters use platinum and
rhodium reactions to convert toxic

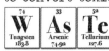

gases from our cars into less harmful gases.

Motor Neurons are the longest cells in the human

running from the lower spinal cord to the big toe.

They Ca N be up to 4.5 feet long.

400 million years ago, a year
was about 400 days long because the

of the Earth on its axis is very gradually slowing down.

This slowing is due to the

87	53	6	22	8	7
Fr	**I**	**C**	**Ti**	**O**	**N**
Francium	Iodine	Carbon	Titanium	Oxygen	Nitrogen
(223)	126.9	12.01	47.87	16.00	14.01

of the tides dragging water back
and forth around the surface.

Fossil fuels such as oil, and natural gas come from sea fossils (trilobites, crinoids and brachiopods) that were alive 570 million years ago...

during the Paleozoic era, long

the dinosaurs.

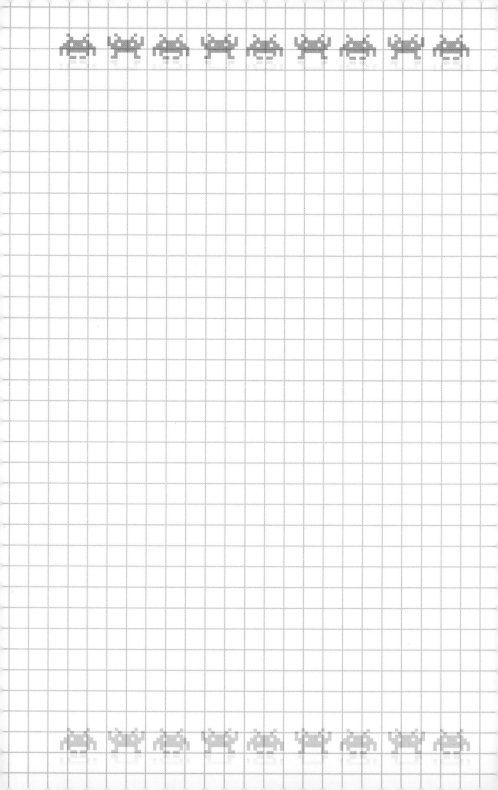

In case you feel old, the Earth is 4.5

years old and

the  is 13.8 billion
years old.

The **B** In **Ar** Y stars of Star Wars'
Tatooine are very much a reality in our Universe.
Check out Kepler-16 and

55 Cancri A. ...but none have been

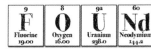

to be habitable. Yet.

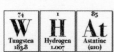 is the name of the first electricity detective?

the nucleus of an atom were the size of a billiard ball…

the entire atom would be the same size
as the Empire State Building in

did the subatomic particle say to the duck?

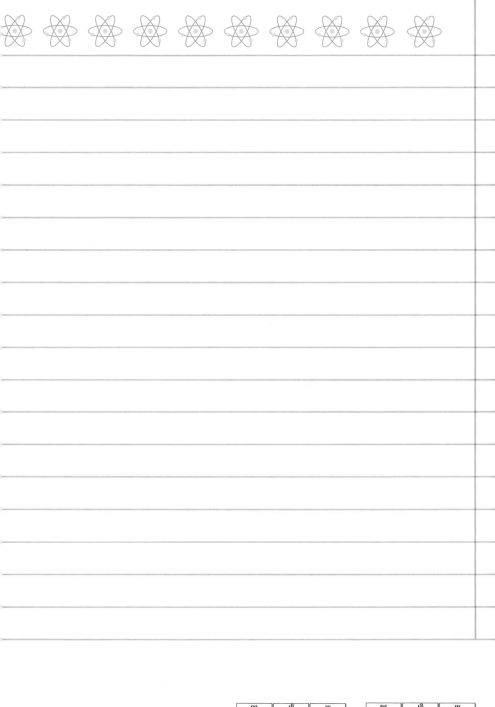

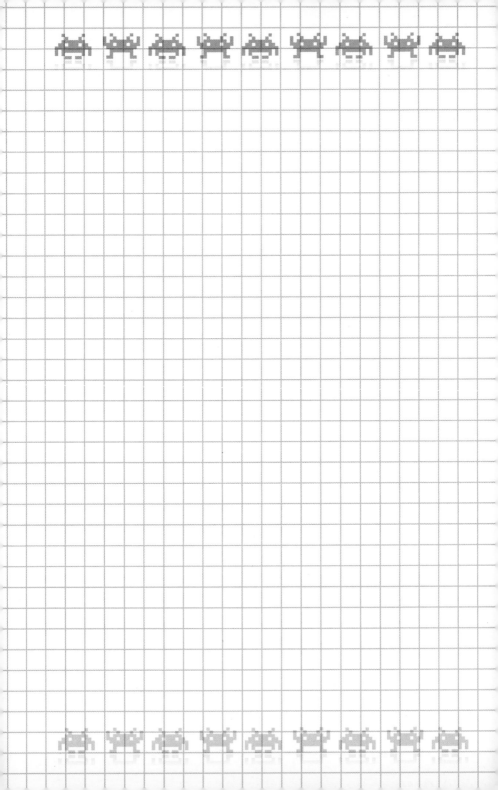

The average lightning bolt is 5 times hotter than the

and each time lightning strikes,

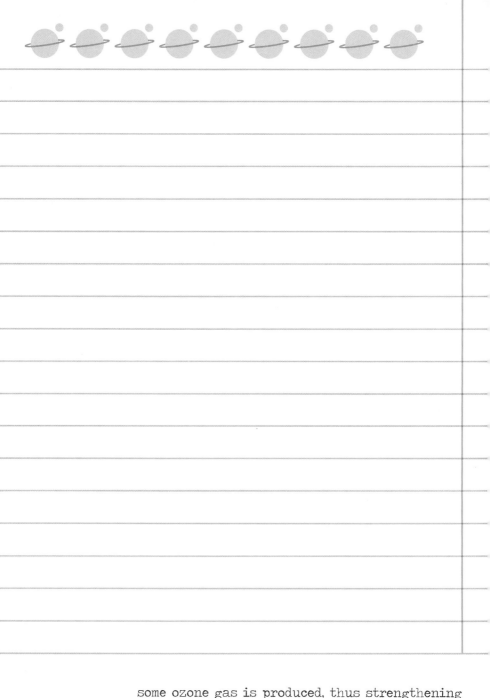

some ozone gas is produced, thus strengthening
the ozone layer in the Earth's

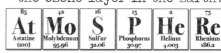

Jupiter Contains two and a

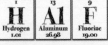

times more mass than the other seven planets combined.

And it rotates in less than ten hours which causes

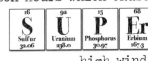

high winds.

Why does a

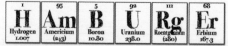

have lower energy than a steak?

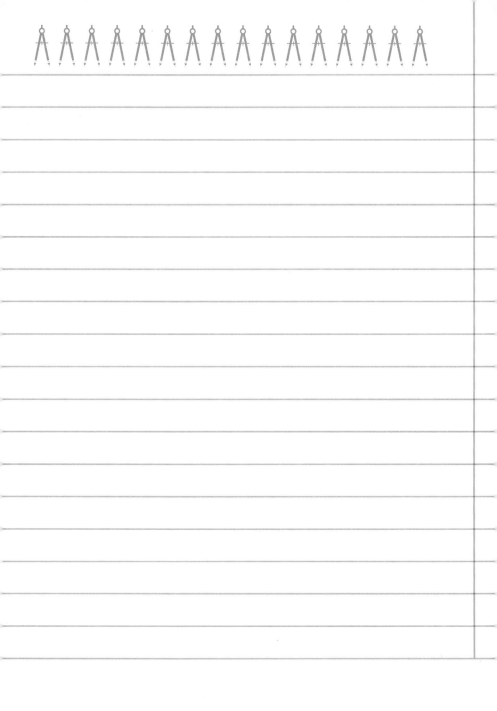

Because it's in a ground

How do you organize a party in **S** **Pa** **Ce** ?

16	91	58
S	Pa	Ce
Sulfur	Protactinium	Cerium
32.06	231.0	140.1

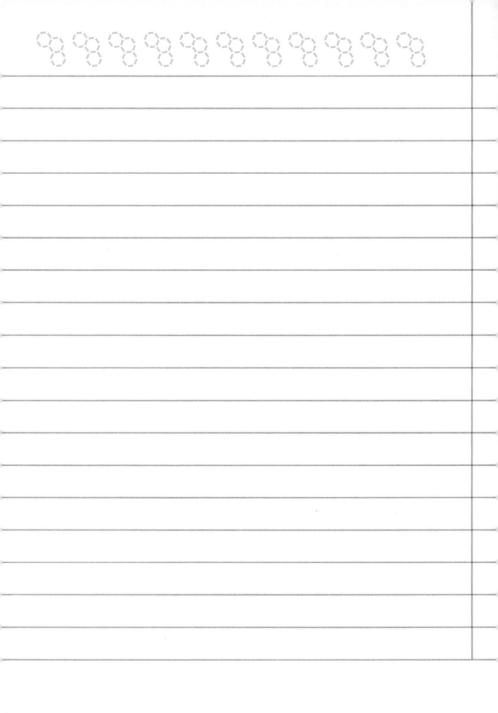

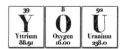

What happened to the man who was stopped for having sodium chloride and a nine-volt in his ?

He
74	33
W	**As**
Tungsten	Arsenic
183.8	74.92

booked for a salt and battery.

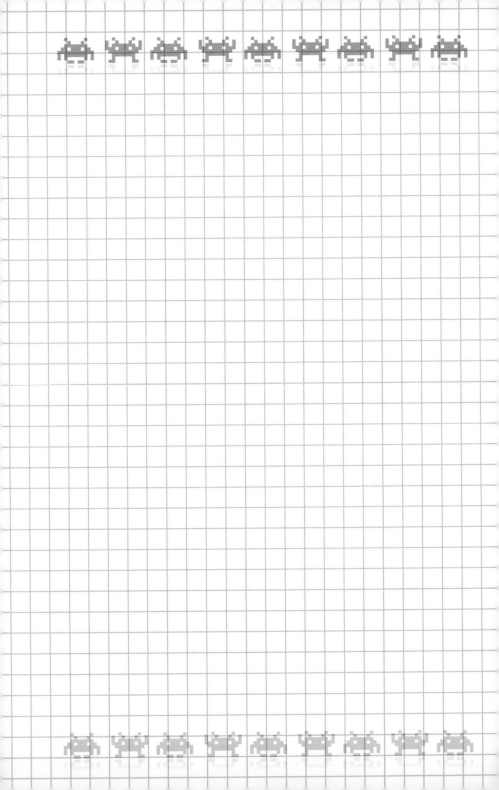

What did the bartender say when oxygen
hydrogen, sulfur, sodoium, and phosphorus
walked into his bar?

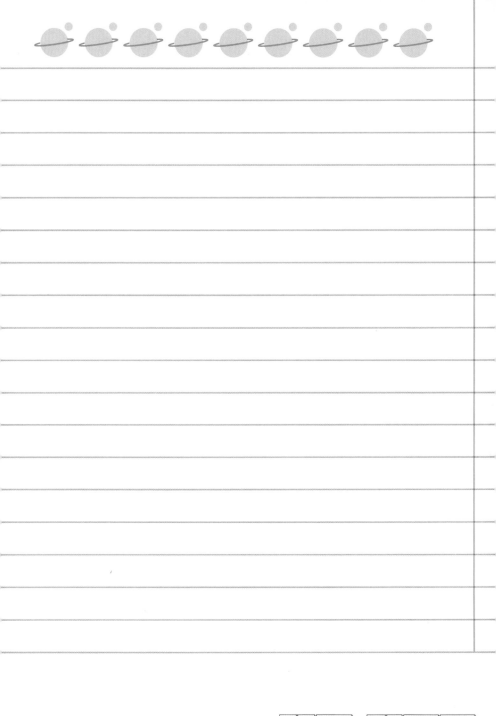

What is the name of 007s

99	19	53	42
Es	**K**	**I**	**Mo**
Einsteinium	Potassium	Iodine	Molybdenum
(252)	39.10	126.9	95.96

cousin?

Polar

5	8	60
B	**O**	**Nd**
Boron	Oxygen	Neodymium
10.80	16.00	144.2

Pericompsus bilbo is a beetle of the

carabidae family, and a South American weevil has
been named Macrostyphlus frodo.

Other species named after Lord of the Rings characters
include an ancient crocodile, a dinosaur and a

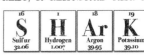

C3PO and R2D2, from the movie Star Wars are close to
being a reality since General Motors and NASA joined

Robonaut 2 (R2), a highly dexterous anthropomorphic robot, is now a permanent resident on the International Space

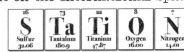

working daily alongside astronauts.

The warning beacons of Gondor, lit by Pippin in
Lord of the Rings, were an ancient form of

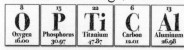

telegraphy...

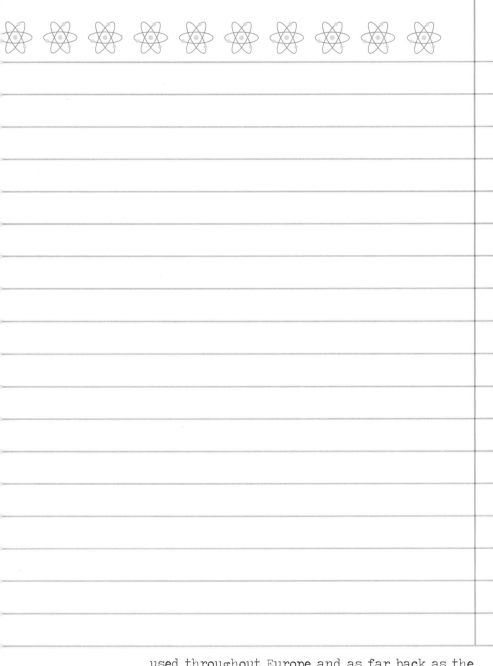

used throughout Europe and as far back as the
Byzantine Empire as a means to communicate and
warn of impending enemy

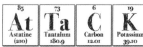

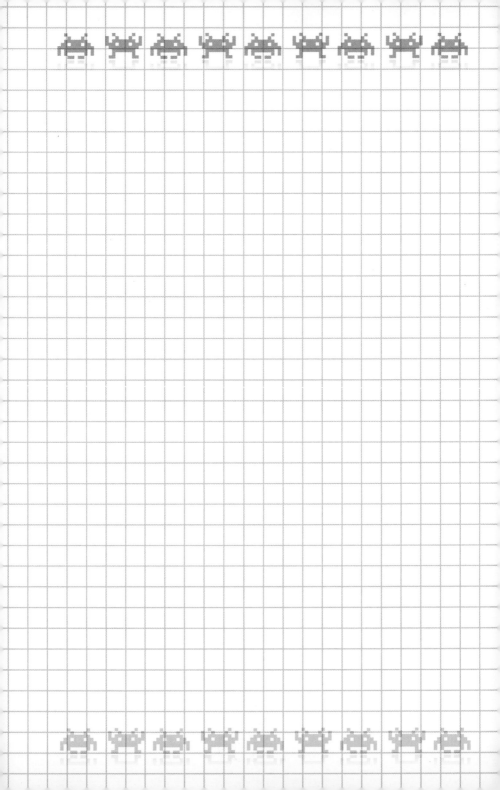

A photon **C** Carbon 12.01 **He** Helium 4.003 **C** Carbon 12.01 **K** Potassium 39.10 **S** Sulfur 32.06 into the hotel. The bellman asks "may I help you with your luggage?"

It **Re** **P** **Li** **Es**

75	15	3	99
Re	**P**	**Li**	**Es**
Rhenium	Phosphorus	Lithium	Einsteinium
186.2	30.97	6.938	(252)

"I don't have any, I'm traveling light."

The communicators used on Star Trek are
somewhat provincial now that cell

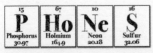

have become so ubiquitous...

and they only shared

| ^{23} V Vanadium 50.94 | ^{68} Er Erbium 167.3 | ^{5} B Boron 10.80 | ^{13} Al Aluminum 26.98 |

information when exploring a strange new world, we
would share videos and tweets.

The most abundant element in the

^{92} U	^{28} Ni	^{23} V	^{68} Er	^{34} Se
Uranium	Nickel	Vanadium	Erbium	Selenium
238.0	58.69	50.94	167.3	78.96

is hydrogen...

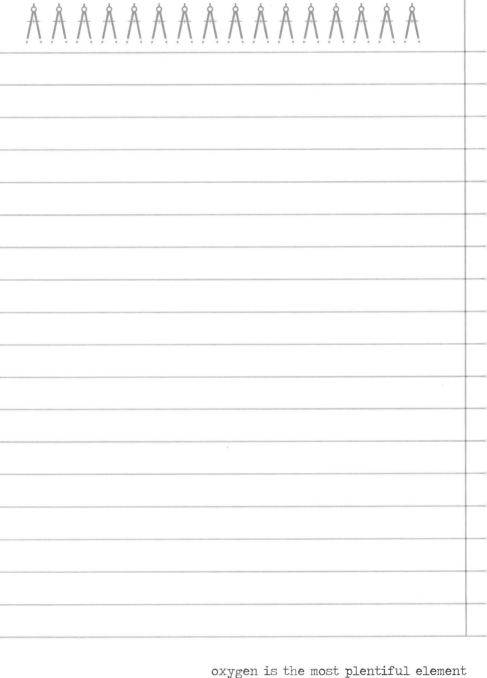

oxygen is the most plentiful element
in the Earth's crust, waters, and

85	42	16	15	2	75
At	**Mo**	**S**	**P**	**He**	**Re**
Astatine	Molybdenum	Sulfur	Phosphorus	Helium	Rhenium
(210)	95.96	32.06	30.97	4.003	186.2

Could Vulcan/human

like our favorite pointy-eared character on Star
Trek ever exist? It seems impossible,

but recombinant DNA has already been utilized
to create interspecies hybrids. Mr.

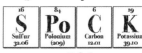

would be proud.

According to the Doppler effect a wave source moving toward YOU will

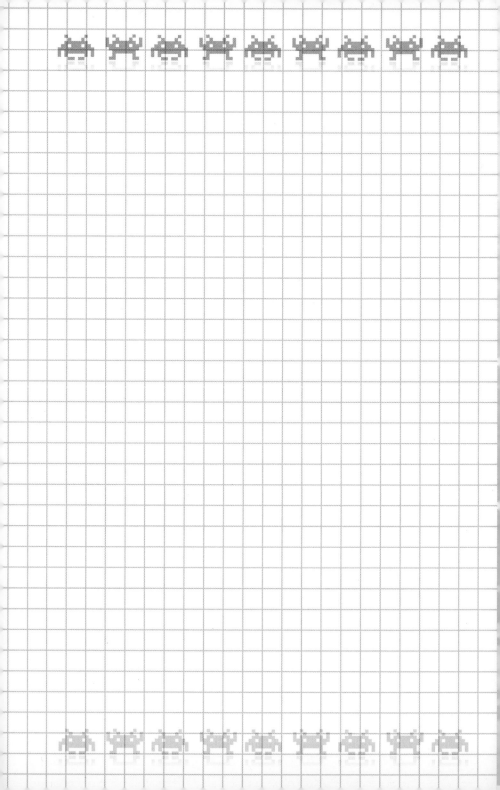

Topographical **Fe** **At** **U** **Re** **S** of the deep ocean Rockall Plateau that lies to the west of Ireland and the UK have been named after famous Tolkien features.

Some examples are the Rohan and
Gondor Seamount, and the

Ridge.

The light saber training droids that Luke used in Star Wars are now a scientific reality, thanks to the brilliant folks at MIT.  are

a volleyball sized satellite, complete with battery

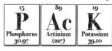

and CO_2 propellant canister that conduct experiments
in the International Space Station laboratory.

The hottest planet in our solar system is

with temperatures of 462° Celsius.

The coldest planet in our solar system is

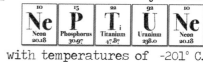

with temperatures of -201° C.

Grand Unified Theories are the theories that

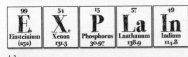

the...

relationship between subatomic particles and the primary

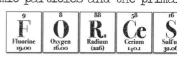

The largest **P Ar Ti C L E** accelerator
is the CERN's Large Hadron Collider in Geneva, Switzerland.
It is capable of accelerating protons to 99 percent the
speed of light while…

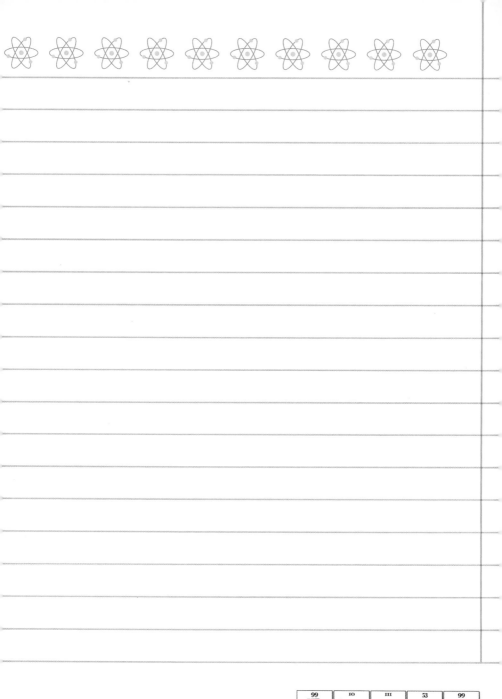

reaching collision
of 14 tera-electron volts and has led to the discovery
of the Higgs boson particle.

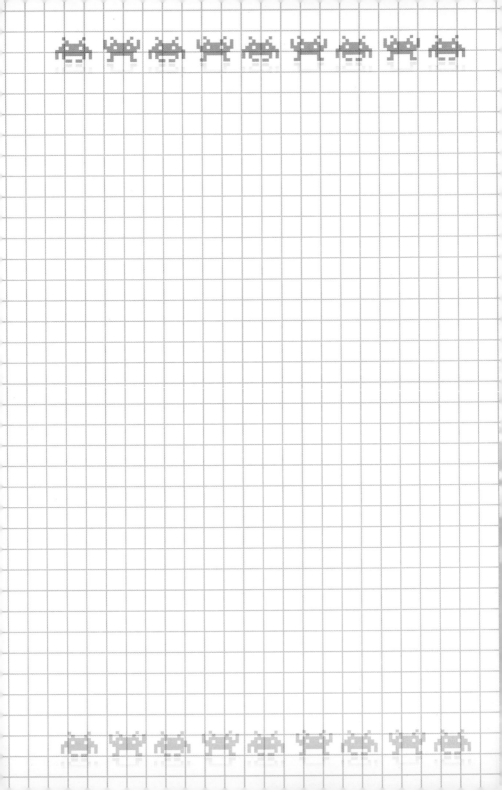

The Eye of Harmony, the engine in Dr. Who's Tardis, is a star on the verge of collapse. According to Einstein's Theory of

75	57	22	23	53	73	39
Re	**La**	**Ti**	**V**	**I**	**T**a	**Y**
Rhenium	Lanthanum	Titanium	Vanadium	Iodine	Tantalum	Yttrium
186.2	138.9	47.87	50.94	126.9	180.9	88.91

it could...

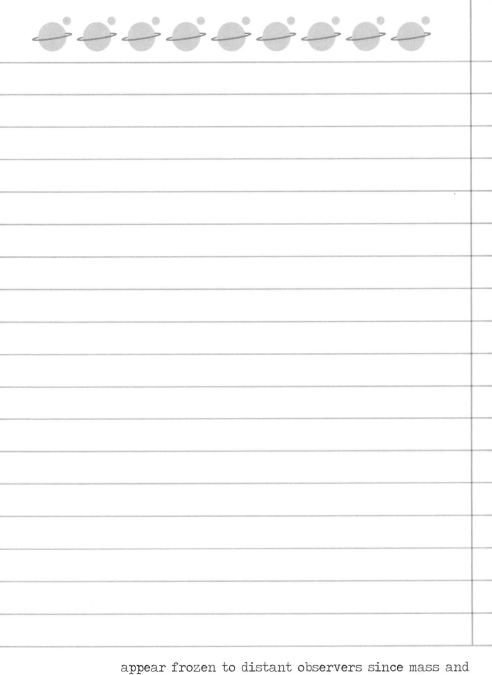

appear frozen to distant observers since mass and

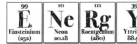

curve space and time, essentially
freezing our view of its collapse.

Astronauts from the Apollo missions staked a laser reflector onto the Moon's surface allowing scientists to periodically shoot a

57	16	68
La	**S**	**Er**
Lanthanum	Sulfur	Erbium
138.9	32.06	167.3

at the Moon and measure the time that it takes to return.

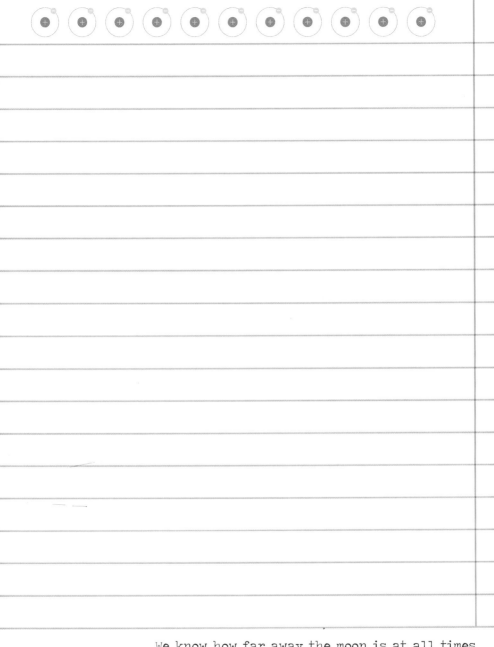

We know how far away the moon is at all times
which silences any skeptics who

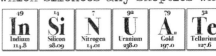

that the moon landings were faked.

The gravitational

energy of an object increases as its
height increases and...

the

19	49	99	22	6
K	**In**	**E**	**Ti**	**C**
Potassium	Indium	Einsteinium	Titanium	Carbon
39.10	114.8	(252)	47.87	12.01

energy of an object changes only
if its velocity changes.

A coulomb is a unit of electric **C** **H** **Ar** **Ge**
an amp is a unit for electric current,
an ohm is a unit of electric resistance and...

Carbon 12.01 Hydrogen 1.007 Argon 39.95 Germanium 72.63

6 1 18 32

a volt is a unit for the difference of potential that
would drive one ampere of current against one ohm of

75	14	16	73	7	58
Re	**Si**	**S**	**Ta**	**N**	**Ce**
Rhenium	Silicon	Sulfur	Tantalum	Nitrogen	Cerium
186.2	28.09	32.06	180.9	14.01	140.1

Or in other words: the desire for the electrons in the
current to move.

Magnetic fields point from the north to the

outside the magnet and

south to north inside the

The time travel of Dr. Who's Tardis into the future
may be possible

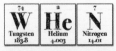

you consider the theories of Albert Einstein.

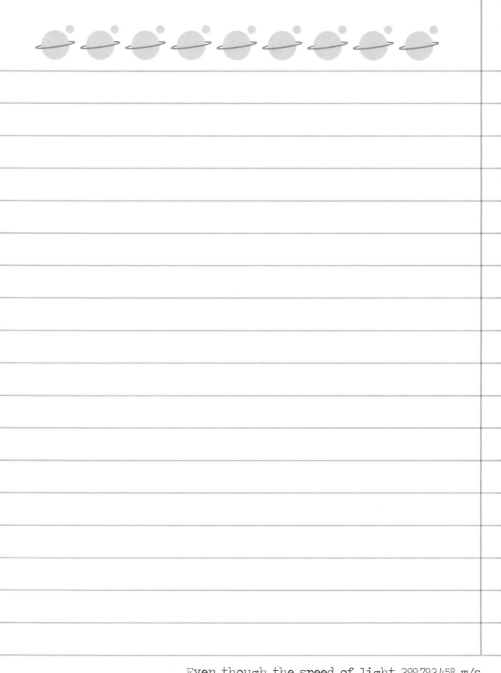

Even though the speed of light 299,792,458 m/s
is a constant, time itself is not an

79	5	16	8	71	52
Au	**B**	**S**	**O**	**Lu**	**Te**
Gold	Boron	Sulfur	Oxygen	Lutetium	Tellurium
197.0	10.80	32.06	16.00	175.0	127.6

Time is dependent on the location of the observer.

7	85	92	88	3
N	**At**	**U**	**Ra**	**L**
Nitrogen	Astatine	Uranium	Radium	Lithium
14.01	(210)	238.0	(226)	6.938

radiation is alpha

Alpha is the least dangerous and gamma is the most

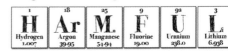

H	Ar	M.	F	U	L.
Hydrogen	Argon	Manganese	Fluorine	Uranium	Lithium
1.007	39.95	54.94	19.00	238.0	6.938

Radio waves and microwaves are

 **B** **O** **Th**

electromagnetic, used for communication, and travel
at the speed of light through space.

Radio waves do not affect organisms, but microwaves do

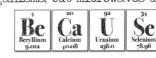

Be	Ca	U	Se
Beryllium	Calcium	Uranium	Selenium
9.012	40.08	238.0	78.96

they cause water and fat molecules to heat up.

The desolation of Mordor, and the Dead Marshes of
Lord of the Rings were heavily

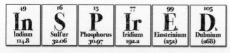

by Tolkien's memories of the

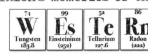

Front of World War I where he was stationed.

44	45	46	47	76	77	78	79
Ru	**Rh**	**Pd**	**Ag**	**Os**	**Ir**	**Pt**	**Au**
Ruthenium	Rhodium	Palladium	Silver	Osmium	Iridium	Platinum	Gold
101.1	102.9	106.4	107.9	190.2	192.2	195.1	197.0

Ruthenium, Rhodium, Palladium, Silver, Osmium, Iridium, Platinum and Gold do not rust.

They are the metals.

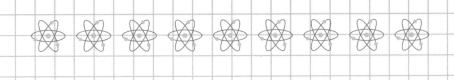

Bibby, Joe. Robonaut. National Aeronautics and Space Administration, 31 May, 2013. Web. January 2015.

Gee, Henry. The Science of Middle Earth: Explaining the Science Behind The Greatest Fantasy Epic Ever Told. CreateSpace Independent Publishing Platform, 2013. Kindle File.

The Science of Dr. Who. Dir. Steve Smith, Ashley Way. Feat. Dr. Brian Cox, 2013, PBS. Film.

Wilson, Jim. The Science of Star Trek. NASA. 15 May, 2009. Web. January 2015. http://www.nasa.gov/topics/technology/features/star_trek

Quarto Publishing Group USA Inc.
142 West 36th Street, 4th Floor
New York, NY 10018

ROCK POINT and the distinctive Rock Point logo are trademarks of
Quarto Publishing Group USA Inc.

ISBN-13: 978-1-63106-087-8

Printed in China

2 4 6 8 10 9 7 5 3 1

www.rockpointpub.com

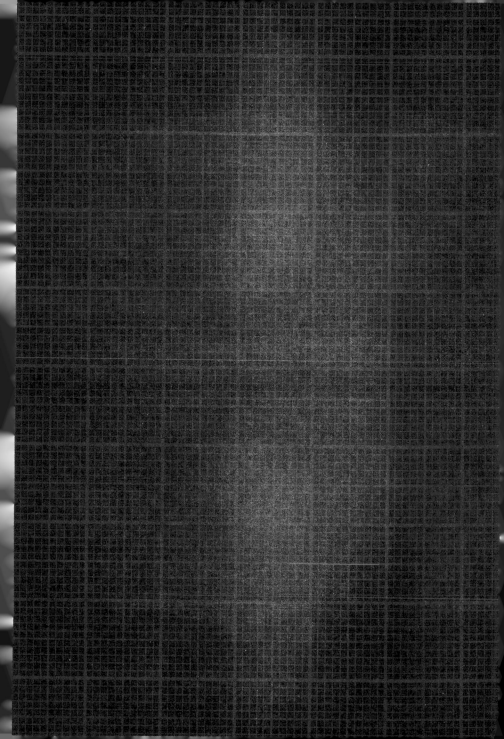